DISCOVER!

EYE ON ECOSYSTEMS

FOOD CHAINS AND FOOD WEBS

THERESA EMMINIZER

Enslow PUBLISHING

Please visit our website, www.enslow.com.
For a free color catalog of all our high-quality books, call toll free
1-800-398-2504 or fax 1-877-980-4454.

Cataloging-in-Publication Data
Names: Emminizer, Theresa.
Title: Food chains and food webs / Theresa Emminizer.
Description: New York : Enslow Publishing, 2023. | Series: Eye on ecosystems | Includes glossary and index.
Identifiers: ISBN 9781978532595 (pbk.) | ISBN 9781978532618 (library bound) | ISBN 9781978532625 (ebook)
Subjects: LCSH: Food chains (Ecology)–Juvenile literature.
Classification: LCC QH541.15.F66 E46 2023 | DDC 577'.16–dc23

Published in 2023 by
Enslow Publishing
2544 Clinton Street
Buffalo, NY 14224

Designer: Tanya Dellaccio
Editor: Theresa Emminizer

Photo credits: Cover Chantal de Bruijne/Shutterstock.com; p. 4 Nynke van Holten/Shutterstock.com; p. 5 Michael Tatman/Shutterstock.com; p. 7 Kadamedo/Shutterstock.com; p. 9 The Len/Shutterstock.com; pp. 10, 17 (fox), 19 (fox), 19 (mouse) Eric Isselee/Shutterstock.com; p. 11 Milan Zygmunt/Shutterstock.com; pp. 13, 17 (eagle), 19 (eagle) FloridaStock/Shutterstock.com; p. 15 Bukhta Yurii/Shutterstock.com; p. 17 (mushrooms) diamant24/Shutterstock.com; pp. 17, 19 (rabbit) Oleksandr Lytvynenko/Shutterstock.com; pp. 17, 19 (clovers) Raras Yulia/Shutterstock.com; p. 19 (snake) Michiel de Wit/Shutterstock.com; p. 19 (bear) Rosa Jay/Shutterstock.com; p. 21 (tetrea fish) Dany Kurniawan/Shutterstock.com; p. 21 (orca) Alexander Baumann/Shutterstock.com; p. 21 (seal) buenaventura/Shutterstock.com; p. 21 (plankton) Napat/Shutterstock.com; p. 21 (algae) JIANG HONGYAN/Shutterstock.com.

Printed in the United States of America

CPSIA compliance information: Batch #CWENS23: For further information contact Enslow Publishing at 1-800-398-2504.

CONTENTS

Boldface words appear in Words to Know.

PASSING ENERGY

An ecosystem is a natural community of living and nonliving things. From the smallest berries to the biggest bears, every member of an ecosystem has a part to play. Food chains and food webs show us how energy flows within an ecosystem.

FOOD BECOMES ENERGY!

PRODUCERS

A food chain starts with producers. Producers are **organisms** such as plants that make their own food without consuming, or eating, other living things. Plants use sunlight for the energy they need to **survive**. In a forest ecosystem, clover is a producer.

PLANTS USE SUNLIGHT FOR ENERGY TO MAKE FOOD THROUGH **PHOTOSYNTHESIS**.

PRIMARY CONSUMERS

Consumers are organisms that survive by eating other living things. A rabbit is a consumer. It eats clover, which is a producer that grows in the forest. Primary consumers are often herbivores, or animals that only eat plants.

RABBITS EAT GREENS AND BERRIES.

SECONDARY CONSUMERS

Secondary consumers eat other consumers. A fox that eats the rabbit is a secondary consumer. Secondary consumers are often carnivores, meaning they eat meat. However, they may also be omnivores, meaning they eat both plants and meat.

THE FOX IS A **PREDATOR**. THE RABBIT IS ITS **PREY**.

TERTIARY CONSUMERS

Tertiary consumers feed on secondary consumers. The top consumers in an ecosystem are called apex predators. In a forest ecosystem, an eagle is an apex predator. It eats the fox that ate the rabbit that ate the clover.

TERTIARY CONSUMERS ARE OFTEN AT THE TOP OF THE FOOD CHAIN.

DECOMPOSERS

Bacteria and **fungi** are decomposers. They make up the last step of the food chain. When plants and animals die, decomposers break down their bodies. This brings **nutrients** into the soil, which helps plants grow. Then the **cycle** begins again.

MUSHROOMS ARE DECOMPOSERS.

TROPHIC LEVELS

Food chains can be formed into trophic levels. An ecosystem's organisms are grouped into different trophic levels based on the type of things they consume. Producers are at the lowest trophic level, primary consumers are in the second trophic level, and so on.

DECOMPOSERS BREAK DOWN ORGANISMS AT EVERY TROPHIC LEVEL.

WHAT'S A FOOD WEB?

There are many food chains within a single ecosystem. Often, these food chains overlap and **connect**, forming a food web. In our example, the food chain was clover-rabbit-fox-eagle. But eagles also eat snakes, making them part of another food chain too.

TWO OR MORE FOOD CHAINS CONNECT TO FORM A FOOD WEB.

USE WHAT YOU KNOW!

Food chains and food webs help us understand how energy flows within the many **habitats** across the planet. What do you think an ocean food chain might look like? What about a desert food chain? Use what you've learned to draw food chains and webs of your own!

HOW WOULD YOU ORDER DIFFERENT PLANTS AND ANIMALS WITHIN THIS FOOD CHAIN?

WORDS TO KNOW

bacteria: Tiny creatures that can only be seen with a microscope.

connect: To join together.

cycle: A repeating series of events.

fungus: A living thing that is somewhat like a plant but doesn't make its own food, have leaves, or have a green color. The plural is fungi. Fungi include molds and mushrooms.

habitat: The natural place where an animal or plant lives.

nutrient: Something living things need to grow and stay alive.

organism: A living thing.

photosynthesis: The process by which green plants and some protists use light to turn water and carbon dioxide into food.

predator: An animal that hunts other animals for food.

prey: An animal that is hunted by other animals for food.

survive: To live through something.

FOR MORE INFORMATION

BOOKS

Krajnik, Elizabeth. *Food Chains in the Forest.* New York, NY: PowerKids Press, 2018.

Regan, Lisa. *Habitats.* New York, NY: PowerKids Press, 2020.

WEBSITES

Britannica Kids
Kids.britannica.com/kids/article/food-chain/353141
Read more about food chains and how they work.

Wild Earth Lab
Wildearthlab.com/2021/08/06/food-chains-and-webs/
Learn more about food chains, food webs, and the players within them.

INDEX